AN INTRODUCTION TO CO-OPERATIVES

A Programmed Learning Text
by Trevor Bottomley and others

Intermediate Technology Publications

Practical Action Publishing Ltd
25 Albert Street, Rugby, CV21 2SD, Warwickshire, UK
www.practicalactionpublishing.com
in association with
Co-operative Educational Materials Advisory Service
of the International Co-operative Alliance

© Intermediate Technology Publications 1979

First published 1979, Reprinted 1987

ISBN 10: 0 903031 63 9
ISBN 13 Paperback: 9780903031639
ISBN Library Ebook: 9781780441696
Book DOI: https://doi.org/10.3362/9781780441696

All rights reserved. No part of this publication may be reprinted or reproduced or utilized in any form or by any electronic, mechanical, or other means, now known or hereafter invented, including photocopying and recording, or in any information storage or retrieval system, without the written permission of the publishers.

A catalogue record for this book is available from the British Library.

The authors, contributors and/or editors have asserted their rights under the Copyright Designs and Patents Act 1988 to be identified as authors of their respective contributions.

Since 1974, Practical Action Publishing has published and disseminated books and information in support of international development work throughout the world. Practical Action Publishing is a trading name of Practical Action Publishing Ltd (Company Reg. No. 01159018), the wholly owned publishing company of Practical Action. Practical Action Publishing trades only in support of its parent charity objectives and any profits are covenanted back to Practical Action (Charity Reg. No. 247257, Group VAT Registration No. 880 9924 76).

Reasonable efforts have been made to publish reliable data and information, but the author and publisher cannot assume responsibility for the validity of all materials or for the consequences of their use.

The manufacturer's authorised representative in the EU for product safety is Lightning Source France, 1 Av. Johannes Gutenberg, 78310 Maurepas, France. compliance@lightningsource.fr

ACKNOWLEDGEMENTS

Financial support for the publication of this series of manuals on co-operatives has been provided by the Ministry of Overseas Development. The Intermediate Technology Development Group gratefully acknowledges their generosity.

Special thanks are due to the Co-operative Department in the Gambia. They arranged a field-test for this book and supplied the additional material which appears now as the white pages of Section 8.

PREFACE

This book is suitable for use by members or prospective members of Co-operatives. It could also be used by people preparing to form a Co-operative. It contains material for five meetings of study groups, including the private study needed before the groups meet, and can also be used by individuals studying alone.

CONTENTS

Introduction 7
1. What is a Co-operative? 11
2. Further explanation of principles 23
3. Types of Co-operatives 33
4. How Co-operatives are organised 45
5. Money and registration 59

INTRODUCTION

Co-operatives can bring prosperity to those who join them, but only if the members now how a Co-operative works. This book has been written to help you to be an active member of a successful Co-operative.

To use the book, work in this way. First read each section in a chapter on your own, testing yourself by completing the sentences at the foot of the section. Then think about the problems on the tinted page at the end of the chapter. If possible, meet with other people who are also studying the book to discuss the problems. Then act on what you have learnt.

Now turn to the next page to find out more about the white study pages.

How to use these white study pages

After reading the first section on a page, complete the sentence you find at the bottom of the section. Then turn over to compare what you have written with the answer given on the back of the page. In this way you will make sure that each section is clear in your mind before you read further.

Complete this sentence

Fill the space in the sentence at the bottom of each section . . . . . . .you turn over.

Answer

Fill the space in the sentence at the bottom of each section *before* you turn over.

Chapter One

WHAT IS A CO-OPERATIVE?

1.1 The word 'co-operate' means 'work together'. A Co-operative is a business formed, owned and controlled by a group of people who agree to follow five special rules in running it. The rules are the PRINCIPLES OF CO-OPERATIVE ORGANISATION and they are so important that the first two chapters of this book will be devoted to explaining them to you.

Complete this sentence

A Co-operative is a .... run on the basis of the .... of
....

1.2 Principles of Co-operative Organisation. Co-operative business organisation is based on five main principles. The first two are:

(a) open and voluntary membership

(b) democratic control.

Complete these sentences

There are .... main .... of Co-operative organisation. The first two are:

(a) .... and ....

(b) ....

1.1 Answer

A Co-operative is a *business* run on the basis of the *Principles* of *Co-operative Organisation*.

1.2 Answer

There are *five* main *principles* of Co-operative organisation. The first two are:

(a) *open* and *voluntary membership*

(b) *democratic control*.

1.3 Principles of Co-operative Organisation. There are five main principles of Co-operative organisation. Two of them have been learnt. The other three are:

(a) LIMITED INTEREST ON SHARE CAPITAL

(b) fair DISTRIBUTION OF SURPLUS (profit)

(c) the promotion of EDUCATION.

Note that for Co-operatives profit is usually called Surplus and interest is sometimes called Dividend.

Complete this sentence by using the words in capitals above

The other three principles refer to ....
.... and ....

1.4 Principles of Co-operative Organisation

Complete this statement

The five main principles of Co-operative organisation are:

(a) .... and ....

(b) ....

(c) .... on ....

(d) .... of ....

(e) the .... of ....

1.3 Answer

The other three principles refer to *limited interest on share capital, distribution of surplus* and *education.*

1.4 Answer

The five main principles of Co-operative organisation are:

(a) *open* and *voluntary membership*

(b) *democratic control*

(c) *limited interest* on *share capital*

(d) *fair distribution* of *surplus*

(e) the *promotion* of *education.*

Each of these will now be discussed in turn, starting with open and voluntary membership.

1.5 Open and voluntary membership. Membership of a Co-operative must be open to all who can benefit from it and members should join voluntarily — of their own free will. This means that unless there is good reason no-one should be prevented from joining a Co-operative if he can benefit from it and will do his duty as a member. It also means that no-one should be forced to join; membership should be voluntary.

Complete this sentence

No-one should be .... from joining a .... if he can benefit from it but no-one should be .... to ....

1.6 Who can benefit? You have just learnt that all who can benefit from membership must be allowed to join a Co-operative. For example, a farmer who needs to market his produce can be a member of a Marketing Co-operative and a housewife who wishes to buy household goods can be a member of a Co-operative which sells the goods she needs. Those who need and can use the services of the Co-operative can be members.

Complete this sentence

People who .... and............. .... the services of a Co-operative can be members.

1.5 Answer

No-one should be *prevented* from joining a *Co-operative* if he can benefit from it but no-one should be *forced* to *join*.

1.6 Answer

People who *need* and *can use* the services of a Co-operative can be members.

1.7 Restricted membership. In certain types of Co-operatives there has to be a restriction on the principle of open membership. For example, if workers form a Co-operative to produce goods together, only those people with skills who can help with production should be members. It is wrong to have more members than there are jobs in this type of Co-operative.

Complete this sentence

A Co-operative for production should have no more members than there are .... and only people with the necessary .... should be allowed to join.

1.8 Unjust restriction of membership. You have learnt that unless there is *good reason* no-one should be prevented from joining a Co-operative if he could benefit from it. What do we mean by good reason? Take an example: a village Farmers' Co-operative must be open to all the farmers in the village or surrounding area. The Co-operative could justly say that only farmers with land in that area could join. It could *not* say that a farmer with land in the area could not join because he was of a different tribe or race or religion or for some other unfair reason.

Complete this sentence

It would always be wrong for a Co-operative to prevent someone joining because of their.......... or .... or ....

1.7 Answer

A Co-operative for production should have no more members than there are *jobs* and only people with the necessary *skills* should be allowed to join.

1.8 Answer

It would always be wrong for a Co-operative to prevent someone joining because of their *tribe* or *race* or *religion*.

1.9 You have finished the first chapter of this book. On the next page you will find the first *Problems for Thought and Discussion.*

Try to work out answers to the problems. Discuss them with other people in your Co-operative and, if possible, get them to read the book too.

If there is a Co-operative officer or other suitable person who is willing to help, send him your solutions to the problems and ask him to comment.

PROBLEMS FOR THOUGHT AND DISCUSSION

1(a) Write down the five Principles of Co-operative Organisation without looking at the earlier pages of this book. If you are working with other people, compare your list with theirs. Then compare with the answer to section 1.4.

1(b) Read sections 1.5, 1.6, 1.7 and 1.8 again and then say in your own words what open and voluntary membership means.

Now answer either 1(c) *or* 1(d)

1(c) If you have not yet joined a Co-operative, describe the work you would like a Co-operative to do for you. What kinds of people should be allowed to be members of a Co-operative for this purpose? What kinds of people should not be allowed to be members?

1(d) Answer this problem if you are already a member of a Co-operative. What kinds of people would benefit from joining your Co-operative? Have all been allowed to join? Is the principle discussed in sections 1.5 to 1.8 being obeyed?

Chapter Two

FURTHER EXPLANATION OF PRINCIPLES

2.1 Democratic control. A Co-operative is owned by the members; it must be run for the benefit of the members. All members are equal. The Co-operative is governed by a general meeting which all members have the right to attend and at which each member has one vote and one vote only. The general meeting will normally elect a committee to supervise the work of the Co-operative. Each member has the right to be nominated for election to the committee and each has one vote in the election.

Complete this sentence

In a Co-operative all members are .... and each has .... only.

2.2 Share capital. Every Co-operative needs some equipment to run its business. For example, most Societies need a safe and an office table. Some need a building and some a lorry. The money used to buy such things is called CAPITAL. It may be possible to borrow some capital but much of it must come from the members. So every member is required to contribute at least a minimum amount of money to the capital of the Co-operative. This money is called SHARE CAPITAL and it remains the property of the member who paid it although it is used by the Co-operative.

Complete this sentence

Each member has to contribute a .... to the .... of the ....

2.1 Answer

In a Co-operative all members are *equal* and each has *one vote* only.

2.2 Answer

Each member has to contribute a *minimum share* to the *capital* of the *Co-operative*.

2.3 Limited interest on share capital. The Co-operative pays members for the use of share capital. For example, members might receive each year an amount equal to 1/20th of the total of their share capital. This amount is called INTEREST or DIVIDEND.

Even if the Co-operative makes a very big surplus, the rate of interest must not rise above a level which is fixed by the members and judged to be fair and reasonable when compared with the rates paid, for example, on a savings deposit at a bank. Because the rate of interest cannot rise above this level, it is said to be limited.

The main purpose of your Co-operative is to give the best possible service to members; it is not to pay unlimited interest on share capital from the surplus.

Complete these sentences

The rate of interest paid on share capital is . . . . to a level fixed by the . . . . It does . . . . rise above this level even if the Co-operative is very successful.

2.4 Distribution of surplus. In a Co-operative business any surplus made belongs to the members and should be used or distributed in such a way as to prevent one member gaining at the expense of others. For this reason the surplus is not distributed entirely on the basis of share capital but in such a way as to be fair to all.

Complete this sentence

It is a principle of Co-operative organisation that any . . . . made belongs to the . . . . and should be . . . . or . . . . in such a way as to prevent one member . . . . at the . . . . of others.

2.3 Answer

The rate of interest paid on share capital is *limited* to a level fixed by the *members*. It does *not* rise above this level even if the Co-operative is very successful.

2.4 Answer

It is a principle of Co-operative organisation that any *surplus* made belongs to the *members* and should be *used* or *distributed* in such a way as to prevent one member *gaining* at the *expense* of others.

Note: We continue to discuss this principle on the next two pages.

2.5 Distribution of surplus (savings or services). There are three main ways of distributing a Co-operative's surplus. Here are two of them:

(a) by using or saving the whole or part to develop the business of the Co-operative;

(b) by providing services for members, such as education.

Complete this sentence

Surplus can be used to .... the business of the Co-operative and to provide services such as ....

2.6 Distribution of surplus (patronage refund). You have learnt two ways in which the surplus of a Co-operative can be used to develop the business and to provide services such as education.

The third way is by distributing surplus to members in proportion to the trade each member has done with the Co-operative. This is sometimes called BONUS or PATRONAGE REFUND. For example, if Elizabeth has bought twice as much as John from the Co-operative this year, her bonus should be twice as big as his.

Complete this sentence

Co-operatives may distribute surplus to members in .... to the .... each member has done with the ....

2.5 Answer

Surplus can be used to *develop* the business of the Co-operative and to provide services such as *education*.

2.6 Answer

Co-operatives may distribute surplus to members in *proportion* to the *trade* each member has done with the *Co-operative*.

2.7 Promotion of education. The fifth principle of Co-operative organisation is the promotion of education. Education is important for two main reasons:

(a) Members must support their Co-operative and use their vote wisely to control it. They therefore need to understand the principles upon which the Co-operative is based and how it works. So education for members is necessary.

(b) The Co-operative is a business serving the economic interests of its members. To do this it has to be run efficiently. So education for committee members and staff is necessary.

Complete this sentence

Co-operative education is important and should be provided for
.........., ... and ....

2.8 Act on what you learn. This now completes the explanation of the five principles which have guided successful Co-operatives in so many parts of the world. It is very important to look for ways of using what you learn to make your Co-operative a good one. The *Problems for Thought and Discussion* which appear on the next page will help you to do this. So will the yellow pages at the end of each later chapter.

Discuss the problems with other people in your Co-operative. Between you, decide what action to take as a result of your discussion. Learning from this book does no good UNLESS YOU ACT ON WHAT YOU LEARN.

Complete this sentence

After reading this book you must .... on....................
....

2.7 Answer

Co-operative education is important and should be provided for *members, committee members* and *staff.*

2.8 Answer

After reading this book you must *act* on *what you learn.*

PROBLEMS FOR THOUGHT AND DISCUSSION

2(a) How can you encourage all the members of your Co-operative to attend meetings and to use their vote for the good of the Co-operative?

2(b) How can you encourage the poorest and least educated of your members to speak at meetings and to stand for election to the committee?

2(c) How much is the minimum share capital each member should contribute to your Co-operative? Have you paid what you should? Have others? Are you helping your Co-operative to give better service by contributing more than the minimum?

2(d) Read sections 2.4, 2.5 and 2.6 again. How should your Co-operative distribute its next surplus?

2(e) Read section 2.7 again to find the three types of people who need education to make a Co-operative successful. What has your Co-operative already done to educate each of these? What education should be provided in the next months? When you answered 2(d) did you remember that some of the surplus should be used for education?

Chapter Three

TYPES OF CO-OPERATIVES

3.1 There are many different types of Co-operatives, each providing a particular service or group of services. People become members of those Co-operatives which provide the services they require. For example, a farmer growing cotton will join a Co-operative which markets cotton. A person who needs a house will join a Housing Co-operative.

The main types of Co-operative will be discussed on the following pages.

Complete this sentence

People become .... of Co-operatives which provide the .... they require.

3.2 Agricultural Credit. An Agricultural Credit Co-operative helps members to obtain loans for better farming. If several farmers join together to borrow as one Co-operative, loans can be obtained more easily and at lower interest.

Lenders are more likely to make loans if they know that the members together will make sure the money is repaid. The loan made to the Co-operative is usually divided up so that individual members receive separate loans.

Complete this sentence

An Agricultural Credit Co-operative helps members to obtain .... for........ ....

3.1 Answer

People become *members* of Co-operatives which provide the *services* they require.

3.2 Answer

An Agricultural Credit Co-operative helps members to obtain *loans* for *better farming*.

3.3 Thrift and Credit (sometimes called Thrift and Loan or Credit Union). A Thrift and Credit Co-operative has two purposes. It encourages members to save regularly, and using the savings thus collected it makes loans to members who need them. In this way it acts as a small Co-operative bank directly controlled by the people (the members) who use it.

Complete this sentence

A Thrift and Credit Co-operative has two purposes: it encourages members to .... and .... to members.

3.4 Consumer. A Consumer Co-operative sells food and household goods and provides services to its members. Since we are all *consumers* anyone can be a member of a Consumer Co-operative. If the goods and services are for use in the members' work, the name Supply Co-operative is generally used and this will be explained in section 3.7.

Complete this sentence

A .... Co-operative sells goods and provides .... to its members.

3.3 Answer

A Thrift and Credit Co-operative has two purposes: it encourages members to *save regularly* and *makes loans* to members.

3.4 Answer

A *Consumer* Co-operative sells goods and provides *services* to its members.

3.5 Housing. There are two main types of Housing Co-operative. In one the Co-operative owns the houses and charges rent to the occupiers who are its members. In the other the Co-operative provides the member with a loan to buy a house. The house belongs to the member, but if the member does not repay the loan, the Co-operative can sell the house to retrieve the money.

Complete this sentence

A Housing Co-operative can either:

(a) .... houses which it .... to ...., or

(b) make .... to .... who wish to buy a house.

3.6 Workers' Co-operatives (also called Production or Industrial Co-operatives). A Workers' Co-operative is an industrial enterprise owned and run by the workers. It might consist of a few people making traditional crafts by hand. It might be a farm, a factory, a construction firm, or any other kind of industrial activity. The essential feature is that the workers in the enterprise are also its owners.

Complete this sentence

A Workers' .... is an .... enterprise .... and .... by the ....

3.5 Answer

A Housing Co-operative can either:

(a) *own* houses which it *rents* to *members,* or

(b) make *loans* to *members* who wish to buy a house.

3.6 Answer

A Workers' *Co-operative* is an *industrial* enterprise *owned* and *run* by the *workers.*

3.7 Marketing and Supply (sometimes called Agricultural). These Co-operatives have two main purposes in service to their farmer members. They organise the marketing of the farmer's products and supply him with the goods and services he needs for working his farm. Members continue to farm separately.

Complete this sentence

The two main purposes of a .... and .... Co-operative are to .... the farmer's .... and .... him with the things he needs.

3.8 Joint Farming. Another type of Agricultural Co-operative is a system of joint farming. The farmers merge their lands for cultivation purposes but retain the ownership of the land and the crop produced. Alternatively, they may become joint owners of agricultural machinery, each using it in turn.

Complete this sentence

In some .... Co-operatives the .... may .... their lands for .... purposes or jointly .... agricultural ....

3.7 Answer

The two main purposes of a *Marketing* and *Supply* Co-operative are to *market* the farmer's *produce* and *supply* him with the things he needs.

3.8 Answer

In some *Agricultural* Co-operatives the *farmers* may *merge* their lands for *cultivation* purposes or jointly *own* agricultural *machinery*.

3.9 Multi-purpose Co-operatives. Most Co-operatives specialise in one service or another: for example, agricultural marketing, consumer goods or housing. Some Co-operatives offer more than one kind of service: for example, agricultural marketing and supply of farm requisites. These Co-operatives which provide several services are called Multi-purpose Co-operatives.

Complete this sentence

Co-operatives which provide .... are called ....

3.10 Types of Co-operatives. There are many different ways in which the co-operative idea can be used. The main types of Co-operatives have been briefly described but there are many more such as insurance, banking, fishing, transport and so on. Most business enterprises can be organised on a co-operative basis. What is common to all is that Co-operatives are owned and controlled by their own members.

Complete this sentence

All Co-operatives are .... and .... by their own ....

3.9 Answer

Co-operatives which provide *several services* are called *Multi-purpose Co-operatives*.

3.10 Answer

All Co-operatives are *owned* and *controlled* by their own *members*.

PROBLEMS FOR THOUGHT AND DISCUSSION

3(a) What are the two main differences between an Agricultural Credit Co-operative and a Thrift and Credit Co-operative? (Sections 3.2 and 3.3 will help you to answer.)

3(b) Read sections 1.7, 3.4 and 3.6 again and then describe the main differences between a Consumer Co-operative and a Workers' Co-operative?

3(c) What is the main difference between a Marketing and Supply Co-operative and a Joint Farming Co-operative? (Sections 3.7 and 3.8 will help you to answer.)

3(d) What is a Multi-purpose Co-operative? (Section 3.9 will help you to answer.)

3(e) What types of Co-operatives would be most suitable for the people of your district?

Chapter Four

HOW CO-OPERATIVES ARE ORGANISED

4.1 Primary Co-operatives. The Co-operatives discussed so far are called PRIMARY Co-operatives. A Primary Co-operative is one which has individual persons as members.

Complete this sentence

A .... Co-operative is one which has individual ....
as ....

4.2 Secondary Co-operatives. A SECONDARY Co-operative is one formed mainly by other Co-operatives which are its members. Secondary Co-operatives are often called Co-operative Unions or Federations. In a Primary Co-operative individual persons are members. In a Secondary Co-operative other Co-operatives are members. For example, several Primary Co-operatives may join together to form a Secondary one called a Co-operative Union.

Complete this sentence

In a Secondary Co-operative other .... are the ....

4.1 Answer

A *Primary* Co-operative is one which has individual *persons* as *members.*

4.2 Answer

In a Secondary Co-operative other *Co-operatives* are the *members.*

4.3 Primary and Secondary Co-operatives. This diagram may help you to remember sections 4.1 and 4.2.

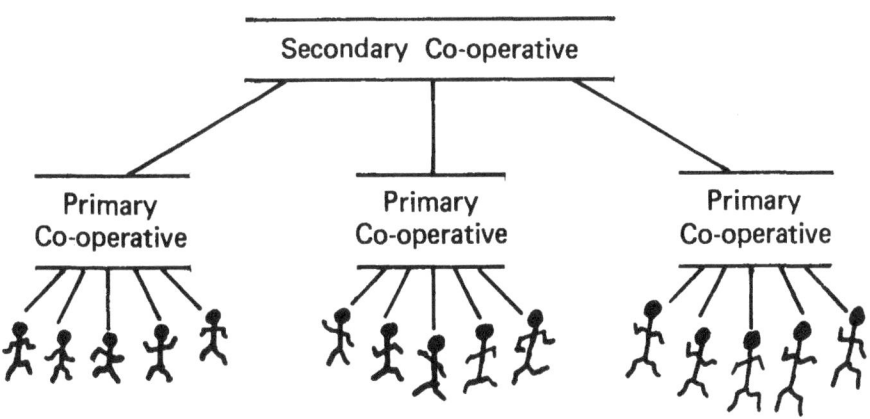

Complete this sentence

.... join Primary Co-operatives and ....
join Secondary Co-operatives.

4.4 Co-operative Unions. A Co-operative Union is a Secondary Co-operative because its members are other Co-operatives. These Unions may include all Co-operatives in one district or region and usually there is one National Union in each country which has for membership all other Co-operatives in the country.

Complete this sentence

A Co-operative Union may cover a .... or .... or a whole ....

4.3 Answer

People join Primary Co-operatives and *Primary Co-operatives* join Secondary Co-operatives.

4.4 Answer

A Co-operative Union may cover a *district* or *region* or a whole *country*.

4.5 Attending meetings and voting. The members own and control their Co-operative. They exercise control through meeting together in a Members' General Meeting where the business of the Co-operative is discussed and decisions are made. Every member can *and should* attend and each has one vote on all matters to be decided.

Complete this sentence

Every member can attend the ....
and each has .... vote on all matters to be decided.

4.6 Annual General Meeting of Members. In most countries the law says that one of the members' general meetings will be the ANNUAL GENERAL MEETING. This important meeting will be held each year. At this meeting the members receive a financial statement about the business done in the past year, discuss plans for the next year and elect a committee.

Complete this sentence

At the .... General Meeting the .... receive a
.... about the business done in the ....
year, discuss .... for the .... year and elect a ....

4.5 Answer

Every member can attend the *Members' General Meeting* and each has *one* vote on all matters to be decided.

4.6 Answer

At the *Annual* General Meeting the *members* receive a *financial statement* about the business done in the *past* year, discuss *plans* for the *next* year and elect a *committee*.

4.7 The Management Committee (sometimes called Board of Directors). It would be difficult for all the members meeting together to supervise efficiently the work of the Co-operative and to make all the necessary decisions. For this reason the members elect from among themselves a MANAGEMENT COMMITTEE. This committee controls the work of the Co-operative on behalf of the members.

The committee must report to the members on how it has run their Society. All members should attend meetings to hear this report and to check that the Co-operative is being well run.

Complete this sentence

The .... Committee .... the work of the .... on behalf of the ....

4.8 The Chairman (or President). One member of the committee is chosen to be its leader and to guide its discussions. He is called the CHAIRMAN or the President. He also guides the discussion at general meetings.

Complete these sentences

The member of the committee chosen to guide its discussions is called the .... He also guides discussion at ....

4.7 Answer

The *Management* Committee *controls* the work of the *Co-operative* on behalf of the *members*.

4.8 Answer

The member of the committee chosen to guide its discussions is called the *Chairman*. He also guides discussion at *General Meetings*.

4.9 The Manager (or Secretary). The Co-operative also requires someone who is responsible for doing the work of the Co-operative. He may be called the MANAGER or the secretary. In a very small Co-operative he may be a member elected to do the work without pay. Usually, however, he is appointed by the committee as a paid worker for the Co-operative. He is responsible to the committee for all the business of the Co-operative.

Complete this sentence

The .... is responsible to the .... for the .... of the Co-operative.

4.10 The manager controls the other workers. The committee should tell the manager what they would like done and then leave him to give the necessary orders to the other people working for the Co-operative. The committee members should not give direct orders to those who work under the manager.

Complete this sentence

.... should not give direct orders to those who work under the manager.

4.9 Answer

The *Manager* is responsible to the *committee* for the *business* of the Co-operative.

4.10 Answer

Committee members should not give direct orders to those who work under the manager.

4.11 A diagram of Co-operative organisation. This diagram shows how Co-operatives are organised.

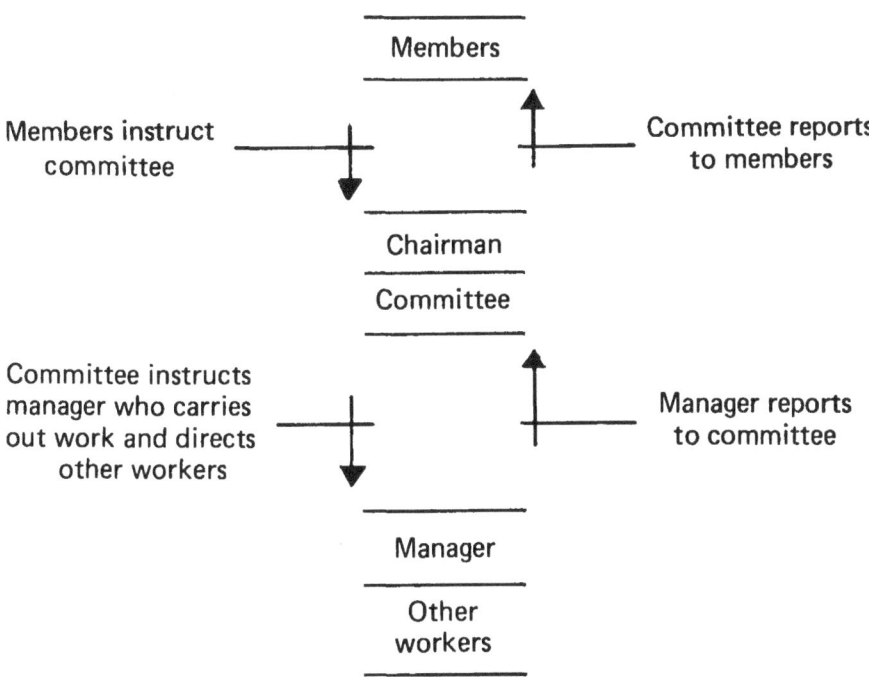

Complete these sentences

The .... instruct the committee. The committee instructs the .... The .... instructs the other workers.

4.11 Answer

The *members* instruct the committee. The committee instructs the *manager.* The *manager* instructs the other workers.

PROBLEMS FOR THOUGHT AND DISCUSSION

4(a) What is the main difference between a Primary and a Secondary Co-operative? Read sections 4.1 and 4.2 if you need help with this question.

4(b) Why is the Members' General Meeting so important in Co-operative organisation? Which Principle of Co-operative Organisation makes General Meetings necessary? Read sections 2.1 and 4.5 if you need help with this question.

4(c) Look again at the diagram in section 4.11. Is your Co-operative organised exactly like this? Are the instructions and reports given just as shown? How should the organisation of your Co-operative be improved?

4(d) Does the Committee give the Manager of your Co-operative clear instructions? Does it interfere too much in his work? How should the relationship between the Manager and the Committee be improved?

Chapter Five

MONEY AND REGISTRATION

5.1 Where does the money come from? A Co-operative is a business. To run a business one needs money. Where does it come from? It can come from three places: from the members, from loans or from the business itself. Now consider each of these in turn.

Complete this sentence

A Co-operative can get .... to run its business from .... places: from the ...., from .... or from the .... itself.

5.2 Money from the members. As discussed in section 2.2, each member has to invest at least the agreed minimum amount of money in the Co-operative when he joins. This money is called SHARE CAPITAL. The members invest it in the Co-operative so that it can be used for business purposes. Usually each member is given a book called a PASS BOOK in which the money he invests in the Society is recorded.

Complete these sentences

Each .... must invest at least the agreed minimum amount of .... in the Co-operative. This money is called ....

5.1 Answer

A Co-operative can get *money* to run its business from *three* places: from the *members,* from *loans* or from the *business* itself.

5.2 Answer

Each *member* must invest at least the agreed minimum amount of *money* in the Co-operative. This money is called *share capital.*

5.3 Money from loans. The share capital may not be enough for running the Co-operative properly and it may be necessary to borrow more. Money borrowed is called LOAN CAPITAL. Money can be borrowed from members or from banks or other places willing to lend. But remember, it is not easy to borrow money. You have to be able to show you can pay it back and you have to pay for borrowing it.

Complete these sentences

Money borrowed is called .... It can be borrowed from .... or .... or other places .... to ....

5.4 Money from business. To run a successful business you have to make money otherwise the business fails. For example, you can buy goods at one price and sell at a higher price. The difference is gross profit or, as it is called in Co-operatives, GROSS SURPLUS.

Complete this sentence

If goods are bought at one price and sold at a .... price, the .... is called .... or ....

5.3 Answer

Money borrowed is called *loan capital*. It can be borrowed from *members* or *banks* or other places *willing* to *lend*.

5.4 Answer

If goods are bought at one price and sold at a *higher* price, the *difference* is called *gross profit* or *gross surplus*.

5.5 The Surplus. From the gross surplus the expenses of running the business, for example wages and rent, have to be paid. What is then left over is called net surplus and belongs to the owners of the business — the members. The members can decide whether all or part of this net surplus should be kept in the business. Using the net surplus to develop the Co-operative's business is the most important way of finding money to run a Co-operative business. You read about it previously in section 2.5.

Complete this sentence

The net surplus can be used to .... the .... of the Co-operative.

5.6 Where does the money come from — a summary. Where the money to run a Co-operative comes from can now be summarised.

Share capital	—	from the members
Loan capital	—	from the members, banks and other bodies willing to lend
Net surplus	—	from the business

Complete this sentence

The money to run a Co-operative comes from ....,
from .... and from ....

5.5 Answer

The net surplus can be used to *develop* the *business* of the Co-operative.

5.6 Answer

The money to run a Co-operative comes from *share capital*, from *loan capital* and from *net surplus*.

5.7 Law. In most countries there is a law called the *Co-operative Law* made by the Government to regulate all Co-operatives. In addition, each Co-operative must have a special set of regulations for itself setting down the way the Co-operative will be organised and controlled. These regulations are called BY-LAWS. The members decide what By-laws are needed, but they must follow the Principles of Co-operative Organisation.

Complete these sentences

The Government makes the .... to regulate all Co-operatives. Members decide on .... for their own Co-operative. By-laws must follow the ....

5.8 Registration. The Government official who administers the Co-operative Law is usually called the Registrar or the Commissioner of Co-operatives. A Co-operative can not operate until the Registrar has agreed that its By-Laws are suitable and that it has obeyed other rules found in the Co-operative Law. The permission given by the Registrar to a new Co-operative to start business is called REGISTRATION. Without registration the members are just a collection of separate people and no Co-operative exists.

Complete these sentences

Without .... there can be no Co-operative. The members are just a collection of separate people. To register, a Co-operative must have suitable ....

5.7 Answer

The Government makes the *Co-operative Law* to regulate all Co-operatives. The members decide on *By-laws* for their own Co-operative. By-Laws must follow the *Principles of Co-operative Organisation.*

5.8 Answer

Without *registration* there can be no Co-operative. The members are just a collection of separate people. To register, a Co-operative must have suitable *By-laws.*

PROBLEMS FOR THOUGHT AND DISCUSSION

5(a) If your Co-operative already exists, find out how much of the money used in its operation came by each of the three ways listed in section 5.6. It is dangerous to depend too much on loan capital so discuss how your Co-operative could obtain more share capital and use more of its surplus for developing its business.

5(b) If you are planning to form a new Co-operative, read section 5.6 again and then discuss how your Co-operative will obtain the money needed to operate. Discuss ways of obtaining as much share capital as possible.

5(c) Which two Principles of Co-operative Organisation are concerned with share capital and the way in which surplus is used? Read sections 2.2 to 2.6 to make sure that you have fully understood these two principles.

5(d) The fifth Principle of Co-operative Organisation is promotion of education (section 2.7). You have been obeying that principle well by studying this book. To make your Co-operative succeed you will need to continue to study. What should be the next stage in your study?

5(e) As a conclusion to your study, make sure that you now know the Principles of Co-operative Organisation by writing down the two that have not yet been mentioned on this page. Read sections 1.5 to 1.8 and also 2.1. If you and other members understand the five principles, your Co-operative has a good chance of success.

www.ingramcontent.com/pod-product-compliance
Ingram Content Group UK Ltd.
Pitfield, Milton Keynes, MK11 3LW, UK
UKHW060455150426
5217IPUK00028B/2082